Dark Matter And Eve Online

Colin Griffith

ISBN 978-0-359-07946-9

Abstract:

This experiment made use of the scanning equipment available to Gallente ships in the multiplayer computer game Eve Online. The purpose of this experiment was to further our understanding of dark matter by using the probe scanner to detect wormholes within the game. The wormholes should be treated like micro-blackholes. It is these micro-blackholes that under the Standard Model could be potential particles of dark matter. Although these wormholes are not literally particles of dark matter, the physics of the game match that of the real world. So our understanding of computer science tells us to assume that these wormholes are found at a level comparable to that of micro-black holes existing within outer space. The comparison between bits of information in a computer and in the real world stands as a testament to this assumption. Although the physical space of a wormhole is not great, indeed the singularity itself would be infinitely small, the mass of one of these micro-blackholes could be great indeed and if they are indeed the best candidate for dark matter they would need to be relatively common in the galaxy (dark matter makes up over a quarter of the universe, significantly more than ordinary matter). Thus, the theory that micro-blackholes are particles of dark matter supports our experiment by analogy.

Procedure:

This experiment was performed using the probe scanner and directional scanner on a Gallente frigate. The analogy is as follows. The sun at the center of each solar system represents an atomic nucleus. The planets orbiting the sun represent electrons and the outer extent of their orbits represents the extent of each electron cloud. The probe scanner serves as an electron microscope and the directional scanner as a radio telescope. Wormholes are the most common type of anomaly and physically they represent a particle in singularity, much like a black hole, with an event horizon that can swallow a ship.

To use the probe scanner, one launches their probes (in this case eight probes were used) into space and then analyzes the results on a diagram of the system. The diagram shows a sphere extending into space around each probe. There are two options for probe formations. The formation can then be moved in its entirety throughout the system. One attempts to drag the probe formation to a location that maximizes the number of spheres surrounding the location of an anomaly, represented by a red x. When the formation is set the pilot performs a scan which gives its result in the form of the probability. For example, locating an anomaly's location to within twenty-five percent. With repeated readings and small adjustments one can increase the probability up to one hundred percent. Once, the scanner reaches seventy-five percent the pilot can fly to the location and observe the anomaly.

The first step in this experiment was to fly to the innermost planet of the system. The distance from that planet to the sun was then recorded. The next step, is to launch the probes and use the technique above to locate an anomaly. Once the anomaly was found, the distance from the ship to the anomaly was recorded. The third step was to fly to the anomaly. The directional scanner was then used to record the distance from the anomaly to the sun. Additionally, the distance from the anomaly to the last planet whose orbit the anomaly is within was also recorded to serve as the distance to the nearest, "electron cloud." Finally, the type of anomaly was also recorded. This experiment was repeated ten times in multiple systems.

Note: All Distances are in astronomical units (AU).

Table 1

Date	System	Planet I to Sun	Distance to Obj	Sun to Object	Type	Within Pl Orbit	Dist. to Orbit Pl	Comment
7/23	Du Annes	0.3	1.68	1.5	Data Site	4	3.3	2 shards
7/24	Grinacanne	0.4	4.54	4.3	Wormhole	4	6.5	
7/25	Renyn	0.3	0.79	0.6	Wormhole	3	1.1	
7/26	Renyn	0.3	4.85	4.9	Wormhole	9	7.4	
	Couster	0.2	3.67	3.8	Gas Site	7	6.3	Acc Gate
	Erne	0.4	3.9	3.6	Wormhole	5	3.2	
	Grinacanne	0.4	1.88	2	Relic	4	4.7	
	Ansone	0.8	7.69	7	Wormhole	6	15.4	
	Ansone	0.8	1.44	1.9	Hideout	2	3	Acc Gate
7/26	Mies	0.4	22.13	21.7	Wormhole	8	2.3	

Analysis

Figure 1

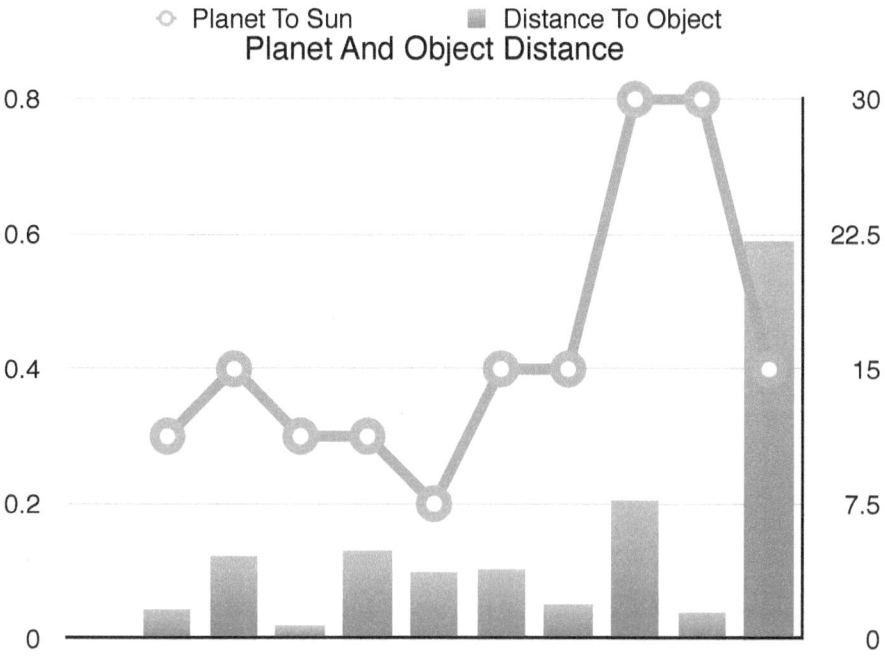

Figure 1 shows the relative patterns of the distance from the first orbiting planet to the sun and the distance from the first orbiting planet it to the object. You can see from the graph that with a few exceptions the shape of each data set is the same. When one graphs goes up or down the other does as well. This is a positive correlation between data sets. It tells us that generally speaking the greater the distance from the planet to the sun, the greater the distance to an anomaly should be.

Figure 2

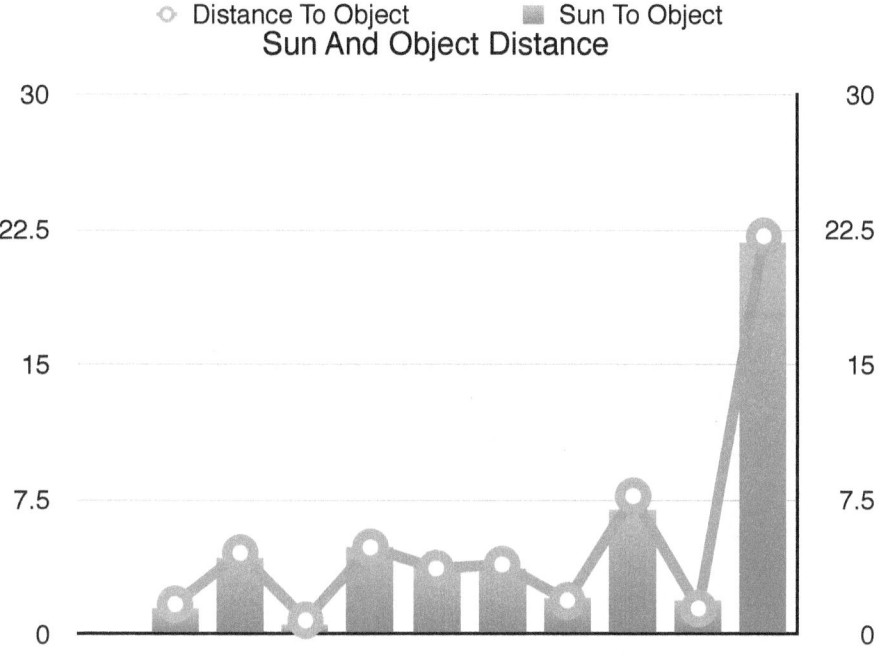

Figure 2 shows the distance to the object and the distance to the subject. You can see that they are almost exactly the same, the distance to the object is just slightly higher than the distance to the sun, but the amount of difference is the same. Thus, there is a one to one correlation between these two distances.

Figure 3

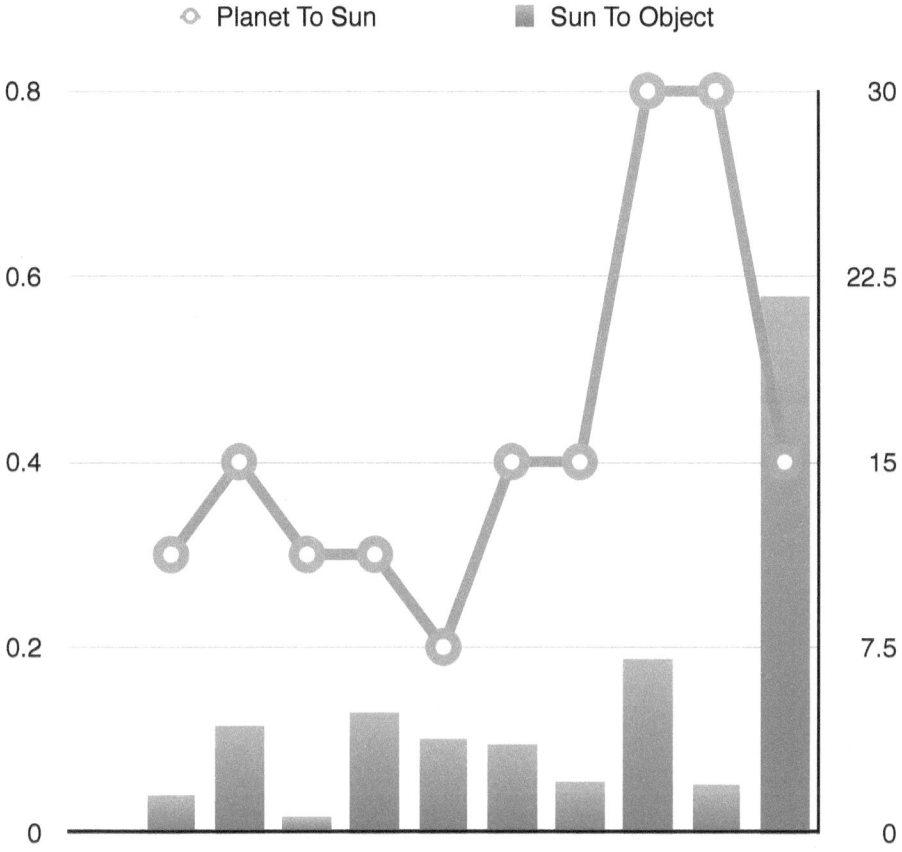

This chart shows the distance from the innermost planet to the sun and the distance from the sun to the object. These two data sets do not appear to be correlated as the shapes of their graphs are very different. It is interesting to note that as the planet to sun distance has its largest decrease the sun to object distance sees its largest increase. This reinforces the fact that there is no correlation here.

Figure 4

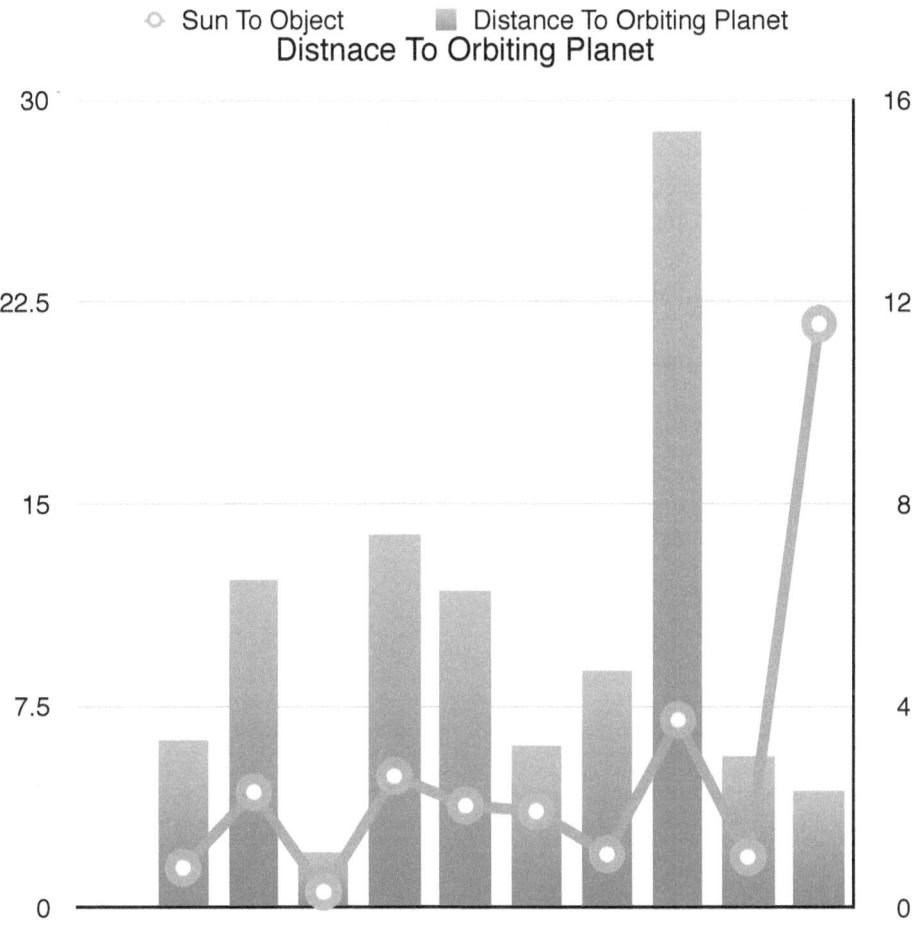

Figure 4 shows the distance from the sun to the object and the distance from the object to the planet whose orbit it lies within. With the exception of the final data point we can see that the shape of the graphs are similar, although there is less variation within the distance from the sun to the object. The final data point appears to be an anomaly itself, although the distance to the orbiting planet sees a small decrease the distance to the orbiting planet sees an enormous increase.

Figure 5

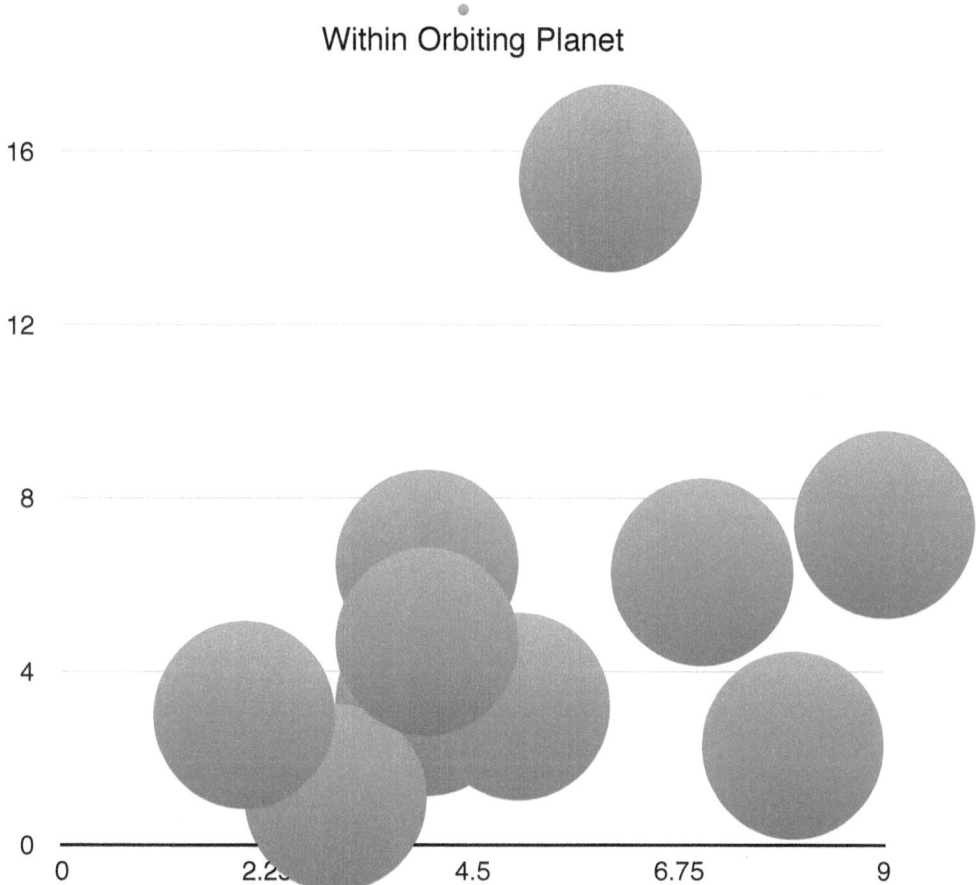

Within Orbiting Planet

This graph shows the position of the innermost planet whose

orbit the anomaly lies within. The x axis indicates the planets

number in the solar system (the numbers increase in order of

distance from the sun). The y axis gives the distance from the

anomaly to this planet. A line of best fit indicates that there is a

positive correlation between these two data sets.

Figure 6

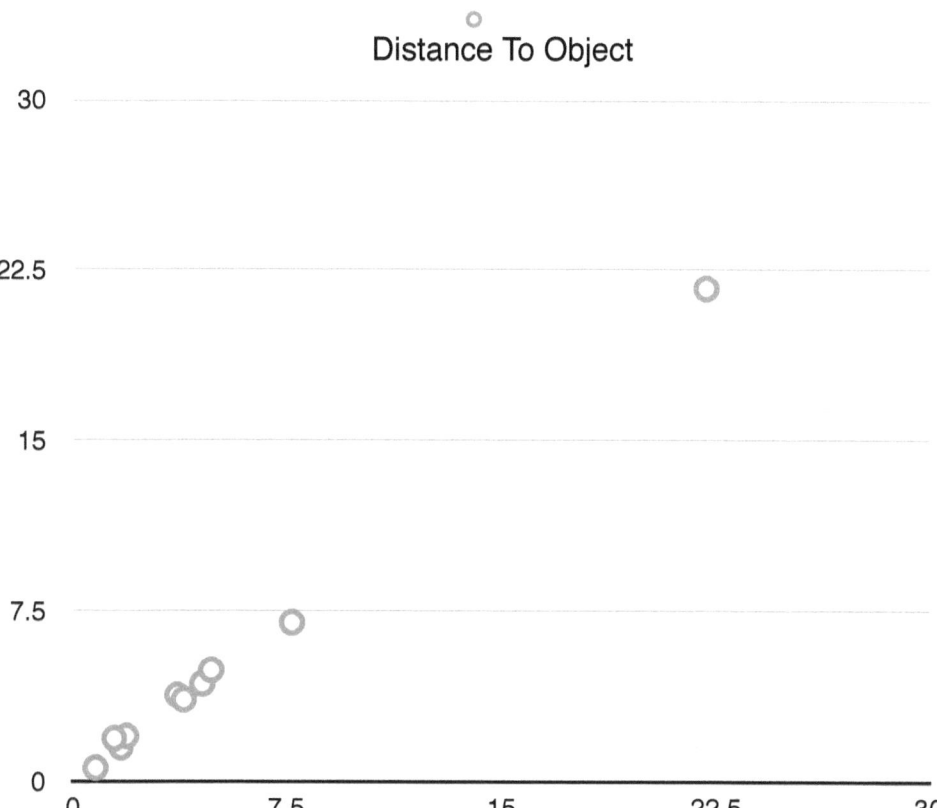

Figure 6 shows the distance to the object against the distance from the sun to the object. The x axis shows the distance to the object. The y axis represents the distance from the sun to the object. The data points form an almost perfectly straight line. We can thus conclude that there is a positive relationship between these two variables. There is a one to one relationship meaning that the two variables increase by the same amount.

Figure 7

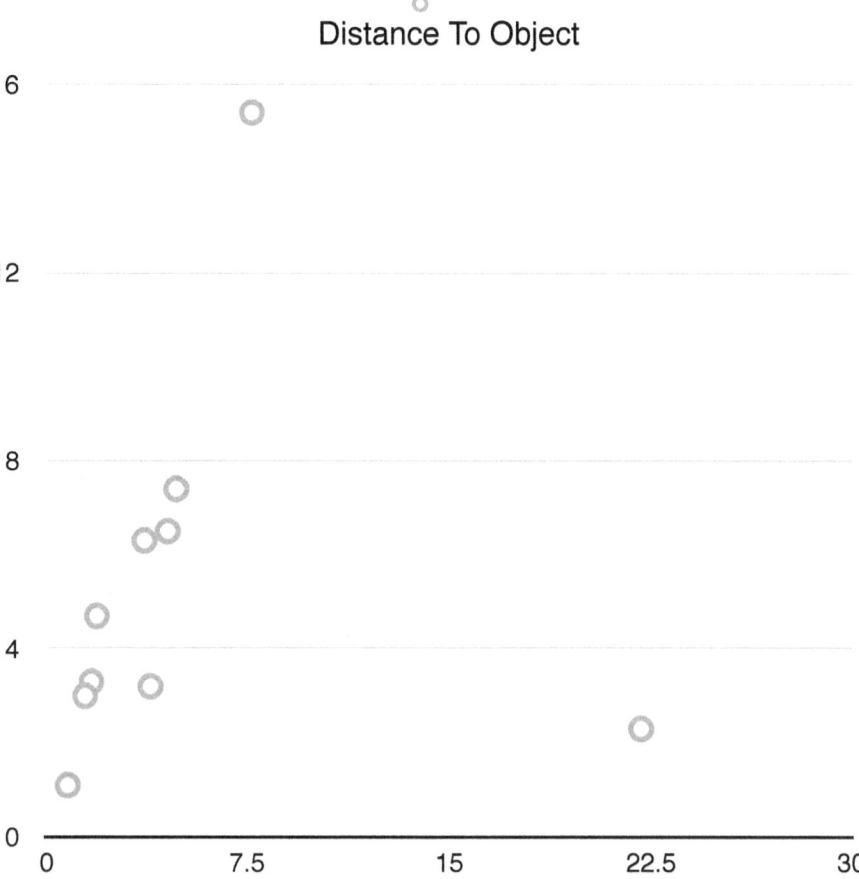

Distance To Object

This chart shows the distance to the object from the innermost planet against the distance from the anomaly to the planet whose orbit it lies within. The x axis shows the distance to the object. The y axis shows the distance from the object to the orbiting planet. With the exception of the last data point, there appears to be a strong correlation between these two variables. As the distance from the object increases the distance to the orbiting planet almost doubles.

Figure 8

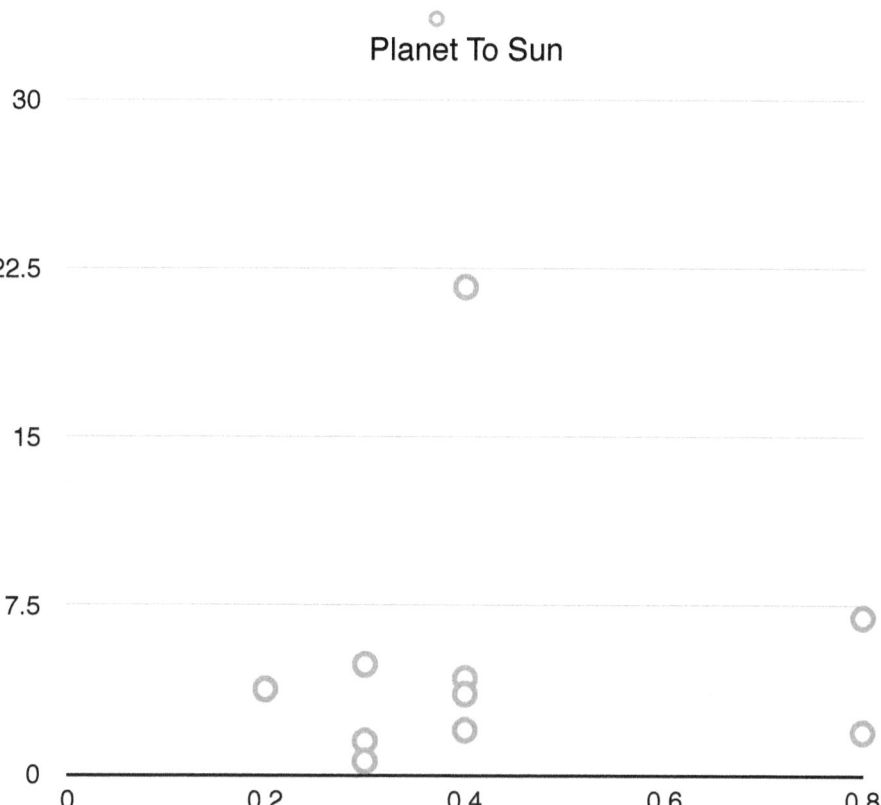

Planet To Sun

Figure 8 shows the distance from the sun to the anomaly

against the distance from the innermost planet to the sun. The

x axis represents the distance from the innermost planet to the

sun. The y axis represents the distance from the sun to the

object. There appears to be no correlation between these two

data sets.

Calculations

These anomalies, and specifically the wormholes, are particles

drifting in three dimensional space with a potential energy

V(x,t). Therefore, we need to find the Schrodinger equation for

a single particle moving in three dimensions in a potential

energy V(x,t). This is problem 4-1 in Feynman's Book,

Quantum Mechanics And Path Integrals, and is found on page

79. We start with a differential equation that relates angular

velocity to time and that incorporates the potential. This

equation is found on page 78 and is equation 4.12. It is the first

equation shown below. We then substitute in the curl for delta

psi over delta x.

$$\frac{\delta\psi}{\delta t} = \frac{-i}{\hbar}\left[-\frac{\hbar^2}{2m}\frac{\delta^2\psi}{\delta x^2} + V(x,t)\psi\right]$$

$$\frac{\delta\psi}{\delta t} = \frac{-i}{\hbar}\left[\frac{-\hbar^2}{2m}\nabla^2 + V(x,t)\psi\right]$$

We then multiply both sides of the equation by psi, distributing psi amongst the variables in the brackets.

$$\frac{\delta \psi^2(x,t)}{\delta t} = \frac{-i}{\hbar}\left[\frac{\hbar^2}{2m}\nabla^2\psi(x,t)+V(x,t)\psi^2(x,t)\right]$$

Finally, we take the square root of both sides and are left with the Schrodinger equation.

$$\frac{\delta \psi(x,t)}{\delta t} = \frac{-i}{\hbar}\left[\frac{-\hbar^2}{2m}\nabla^2\psi(x,t)+V(x,t)\psi(x,t)\right]$$

The complex conjugate function psi* satisfies the equation below. This is shown in Problem 4-3 on page 80.

$$\frac{\delta \psi^*}{\delta t} = \frac{i}{\hbar} (H\psi)$$

First, we insert the Hamiltonian and expand the equation. Then, we change every i to a negative i. We then divide both sides of the equation by -i and multiply by Plank's constant.

$$\frac{\delta \psi^*}{\delta t} \cdot \frac{\hbar}{i} = \left[\frac{1}{2m} \left(\frac{-\hbar}{i} \nabla - \frac{e}{c} A \right) \left(\frac{-\hbar}{i} \nabla - \frac{e}{c} A \right) \psi^* + e\phi\psi^* \right]$$

We then replace the Hamiltonian with negative H.

$$\frac{\delta\psi^*}{\delta t} - \frac{\hbar}{i} = -H\psi^*$$

Finally, we multiply both side by i and divide by Plank's constant to get

$$\frac{\delta\psi^*}{\delta t} = \frac{i}{h}H\psi^*$$

It is important to include the functional derivative. This will let us know how small changes in the argument function, in this case the Schrodinger equation for a free particle, will effect the end result. According to Feynman, "The functional F[x(t)] gives a number for each function x(t) that we may choose." (Quantum Mechanics and Path Integrals, 170) We begin with the equation:

$$S[x(t)] = \int_{t_a}^{t_b} L(\dot{x}, x, t) \, dt$$

This equation is found in problem 7-1 on page 171 of <u>Quantum Mechanics And Path Integrals</u>. We need to show that for any s inside the range t of a to t of b where partial derivatives are evaluated at t=s

$$\frac{\delta S}{dx(s)} = -\frac{d}{ds}\left(\frac{\delta L}{\delta \dot{x}}\right) + \frac{\delta L}{\delta x}$$

Next, we differentiate both sides of our original integral equation and write the following equation. We then divide both sides of the differential equation by ds.

$$\delta S = -\left(\frac{\delta L}{\delta \dot{X}} + \frac{\delta L}{\delta x} + \frac{\delta L}{s} \right) ds$$

$$\frac{\delta S}{ds} = -\left(\frac{\delta L}{\delta \dot{x}\, ds} + \frac{\delta L}{\delta x ds} \right)$$

This last equation simplifies to our goal.

$$\frac{\delta S}{dx(s)} = -\frac{d}{ds}\left(\frac{\delta L}{\delta \dot{x}} \right) + \frac{\delta L}{\delta x}$$

Thus, we have shown that the variable S, our functional,

depends on position (x) which depends on time (t) and it is

related by the integral of the Lagrangian.

Conclusion

Let me summarize my findings as thus. First, we have found that sixty percent of the located system anomalies were wormholes. It is these wormholes that we are interested in as they are candidate for dark matter. Second, we have found that there is a positive relationship between the number of planets in the system and the order of the innermost planet whose orbit the anomaly lies within. This tells us that the greater the number of planets in a system, the greater the probability that the orbiting planet will be farther out in the solar system.

Next, we have discovered a positive, one to one relationship between the distance to the anomaly and the distance from the sun to the anomaly. This means that the distance from the sun to the innermost planet is a factor in how distant the anomalies are.

Additionally, we have found that that there is a positive 2:1 relationship between the distance from the innermost planet in the system to the anomaly and the distance from the anomaly to its orbiting planet. With the exception of one outlier,

32

we see that as the former distances increases by one the latter increases by two. Hence, we have learned that the distance from the innermost planet is a factor in the distance from the anomaly to the orbiting planet. This is important because if the planet is an electron by analogy, we need to know the position of the anomaly relative to it. Since the most common anomaly is a wormhole (dark matter particle) having derived a relationship between these two variables tells us that the position of the first planet (electron) orbiting the sun (nucleus) affects the position of the wormholes (dark matter particles). This is a stunning finding that should help in locating real dark matter particles.

Finally, we have established that the wormhole is a quantum singularity that represents a dark matter particle. This dark matter particle is a free particle moving about in a potential energy. We have derived the Schrodinger Equation for a free particle moving in a potential $V(x,t)$. This equation describes the motion of the dark matter particle within the system (atom). Additionally, we have used the complex conjugate function to show that psi (the position of the particle) depends on the

Hamiltonian. Lastly, we have used functional derivatives to show how changes in the equation affect changes in the position of the particle.

We have learned much about the existence of wormholes in the solar system, and of micro-blackholes orbiting nuclei. Thus, after having learned so much I have convinced myself that micro-blackholes are the best candidate for particles of dark matter. These particles should exist in great abundance and I hope that science will soon confirm their existence in the real world.

Works Cited

Feynman, Richard Phillips, et al. Quantum Mechanics and Path

Integrals. Dover Publications, 2014.

www.ingramcontent.com/pod-product-compliance
Lightning Source LLC
Chambersburg PA
CBHW020958180526
45163CB00006B/2412